Developing Numeracy
MEASURES, SHAPE AND SPACE
ACTIVITIES FOR THE DAILY MATHS LESSON

year
1

Ann Montague-Smith

A & C BLACK

Contents

Reprinted 2002, 2003
Published 2001 by A & C Black Publishers Limited
37 Soho Square, London W1D 3QZ
www.acblack.com

ISBN 0-7136-5876-2

Copyright text © Ann Montague-Smith, 2001
Copyright illustrations © Martin Pierce, 2001
Copyright cover illustration © Charlotte Hard, 2001
Editors: Lynne Williamson and Marie Lister

The author and publishers would like to thank Madeleine Madden and Corinne McCrum for their advice in producing this series of books.

A CIP catalogue record for this book is available from the British Library.

Printed in Great Britain by Caligraving Ltd, Thetford, Norfolk.

A & C Black uses paper produced with elemental chlorine-free pulp, harvested from managed sustainable forests.

Introduction

Developing Numeracy: Measures, Shape and Space is a series of seven photocopiable activity books designed to be used during the daily maths lesson. They focus on the fourth strand of the National Numeracy Strategy *Framework for teaching mathematics*. The activities are intended to be used in the time allocated to pupil activities; they aim to reinforce the knowledge, understanding and skills taught during the main part of the lesson and to provide practice and consolidation of the objectives contained in the framework document.

Year 1 supports the teaching of mathematics by providing a series of activities which develop essential skills in measuring and exploring pattern, shape and space. On the whole the activities are designed for children to work on independently, although due to the young age of the children, the teacher may need to read the instructions with the children and ensure that they understand the activity before they begin working on it.

Year 1 encourages children to:

- use and understand the language of measure, pattern, shape, space and time;
- estimate, measure and compare lengths, masses and capacities, and to suggest suitable units and equipment for such measurements;
- understand the passing of time and to begin to tell the time;
- use everyday language to describe features of familiar 3-D and 2-D shapes;
- recognise and repeat patterns using 3-D and 2-D shapes;
- use a variety of shapes to make models, pictures and patterns, and describe them;
- use everyday words to describe position, direction and movement.

Extension

Many of the activity sheets end with a challenge (**Now try this!**) which reinforces and extends the children's learning, and provides the teacher with the opportunity for assessment. Again, it may be necessary to read the instructions with the children before they begin the activity. For some of the challenges the children will need to record their answers on a separate piece of paper.

Organisation

Very little equipment is needed, but it will be useful to have available coloured pencils, counters, scissors, glue, dice, cubes, small mirrors, small clocks with moveable hands and solid shapes. You will need to provide sets of objects as shown for page 15, a variety of fruit and vegetables for page 16, and one-minute sand timers for page 32.

The children should also have access to a metre stick, a balance with objects to balance, and different containers for filling and pouring, to give them practical experience of length, mass and capacity.

Some of the sheets require cutting and sticking; this can be done by the children or an adult as appropriate.

To help teachers to select appropriate learning experiences for the children, the activities are grouped into sections within each book. However, the activities are not expected to be used in that order unless otherwise stated. The sheets are intended to support, rather than direct, the teacher's planning.

Some activities can be made easier or more challenging by masking and substituting some of the numbers. You may wish to re-use some pages by copying them onto card and laminating them, or by enlarging them onto A3 paper.

Teachers' notes

Very brief notes are provided at the foot of each page giving ideas and suggestions for maximising the effectiveness of the activity sheets. These can be masked before copying.

Structure of the daily maths lesson

The recommended structure of the daily maths lesson for Key Stage 1 is as follows:

Start to lesson, oral work, mental calculation	5–10 minutes
Main teaching and pupil activities *(the activities in the **Developing Numeracy** books are designed to be carried out in the time allocated to pupil activities)*	about 30 minutes
Plenary *(whole-class review and consolidation)*	about 10 minutes

Whole-class warm-up activities

The following activities provide some practical ideas which can be used to introduce or reinforce the main teaching part of the lesson.

Measures

Comparing length

From a collection of strips of paper, ask three children to each choose a strip. They fix their strips to a flip chart so that the other children can see them. Ensure that the three strips are placed with one end level for direct comparison. Ask: *Which is the longest/shortest/ widest/narrowest?* Repeat for another set of three strips, then extend to four.

Comparing mass

Provide three items which will fit into the buckets of a bucket balance. Ask a child to choose two, and to place one in each bucket. Ask: *Which is heavier/lighter? How do you know that?* Encourage the children to observe the bucket balance and to explain what the position of the buckets means.

Now ask a child to replace one of the objects with another. Ask the children to order the objects by weight. Repeat for another set of three objects, then extend to four.

Comparing capacity

You will need a transparent straight-sided beaker with a rubber band around its mid-point; sand or coloured water; and a scoop for filling. Ask the children to estimate how many scoops of sand are needed to fill the beaker to the band, then ask a child to carry this out. Discuss how accurate their estimates were. Move the rubber band to another position and repeat the activity. Talk about how, with practice, estimation becomes more accurate.

Telling the time

Set the hands of a teaching clock to 8 o'clock and ask: *What time is it?* Repeat for other 'o'clock' times, and, when the children are confident, introduce 'half past'.

Set the clock to 6 o'clock and ask: *What time will it be in one hour… two hours… half an hour?* Repeat for other times, and show the passing of time on the clock for the children to check the new time.

In one minute

Ask the children to write their name neatly as many times as they can in one minute. Repeat for other tasks.

Ask the children to shut their eyes for what they think is one minute. Discuss how accurate their estimations were and how they calculated the passing of time.

Shape and space

Children patterns

During a PE lesson, ask the children to work in groups of six or eight and to invent their own repeating pattern in a straight line. They might stand in a line with hands above their heads, then by their sides, alternating along the line. Ask the groups to copy each others' patterns.

Now ask the children to stand in a circle in their group and to invent a pattern that repeats around the circle so that the pattern is complete. Again, invite the children to copy other groups' patterns.

Behind the wall shapes

Put together a selection of 3-D shapes (cubes, cuboids, cylinders, pyramids, cones and spheres) **or** 2-D shape tiles (squares, rectangles, triangles, circles, stars, crescent moons). Gradually slide a shape up above a screen (a piece of card or a book) so that it just peeks over the top. Ask: *What shape might this be? Why do you think that?* Keep showing a little more of the shape until the children work out which shape it is and can name it correctly.

Mystery shapes

Put a selection of 3-D shapes or 2-D shape tiles inside a feely bag, letting the children see what has gone into the bag. Now explain that you will put your hands in the bag and choose one shape. The children ask questions about the shape, to which you can answer only *yes* or *no*. For a 3-D shape they might ask: *Does it have straight edges/points/square faces?*

PE positions, movements and directions

Ask the children to stand in a certain position, such as in front of, behind, next to or opposite a partner. Check that they understand the language and can follow the instructions. Repeat for other positions.

Ask the children to give/follow instructions, for example: *walk forwards/backwards five paces; turn left/right; face the front; make a half/full turn, go behind the bench.*

Turtle moves

Using programmable toys, such as Roamer or Pip, ask the children to suggest how to programme the toy to move it from one point to another. Encourage them to use the language of movement and direction, such as: *Send it forward 5; turn it left 2.*

Alien world

• **Draw the missing alien.**

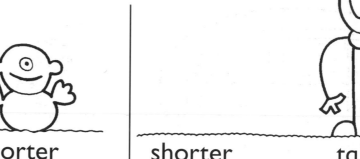

| taller | shorter | shorter | taller |

• **Draw the missing spaceship.**

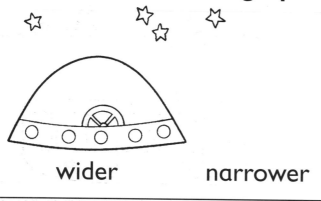

| wider | narrower | wider | narrower |

• **Draw the missing alien pet.**

longer  shorter

shorter longer

Now try this!

• **Draw 3 aliens in order of** boxed(height). **Start with the shortest.**

Teachers' note Provide a range of objects to compare. Discuss the differences between longer/shorter/taller and wider/narrower. Encourage the children to use comparative language when making direct comparisons of length.

Developing Numeracy Measures, Shape and Space Year 1 © A & C Black

Out and about

- **Ring the** longest .

- **Ring the** shortest .

- **Draw the** longest **object in the classroom.**
- **Draw the** shortest **object in the classroom.**

Teachers' note Encourage the children to describe each picture and to use comparative language to compare the pictures in each set, for example: 'The lorry is longer than the car; the bus is the longest.'

**Developing Numeracy
Measures, Shape and Space
Year 1**
© A & C Black

Dinosaurs rule!

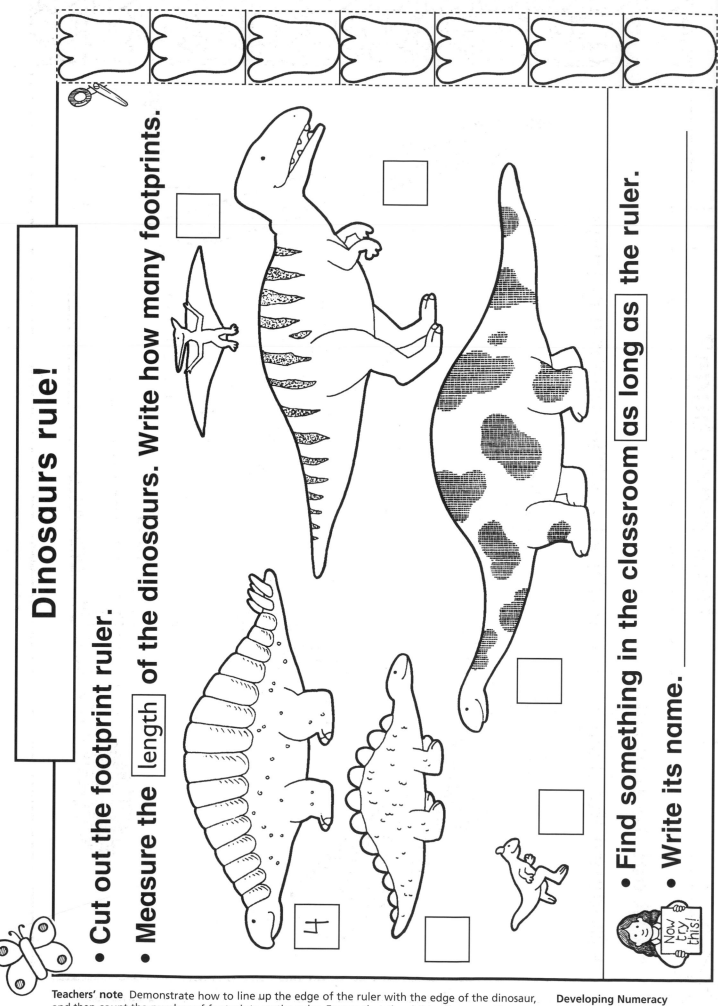

- Cut out the footprint ruler.

- **Measure the** length **of the dinosaurs. Write how many footprints.**

- Find something in the classroom as long as the ruler.

- Write its name.

Teachers' note Demonstrate how to line up the edge of the ruler with the edge of the dinosaur, and then count the number of footprints on the ruler. Ensure that the children measure the length of the dinosaur, not the height.

Developing Numeracy
Measures, Shape and Space
Year 1
© A & C Black

How many bugs?

- **Cut out the bug ruler.**
- **Measure the** length **of the objects. Write how many bugs.**

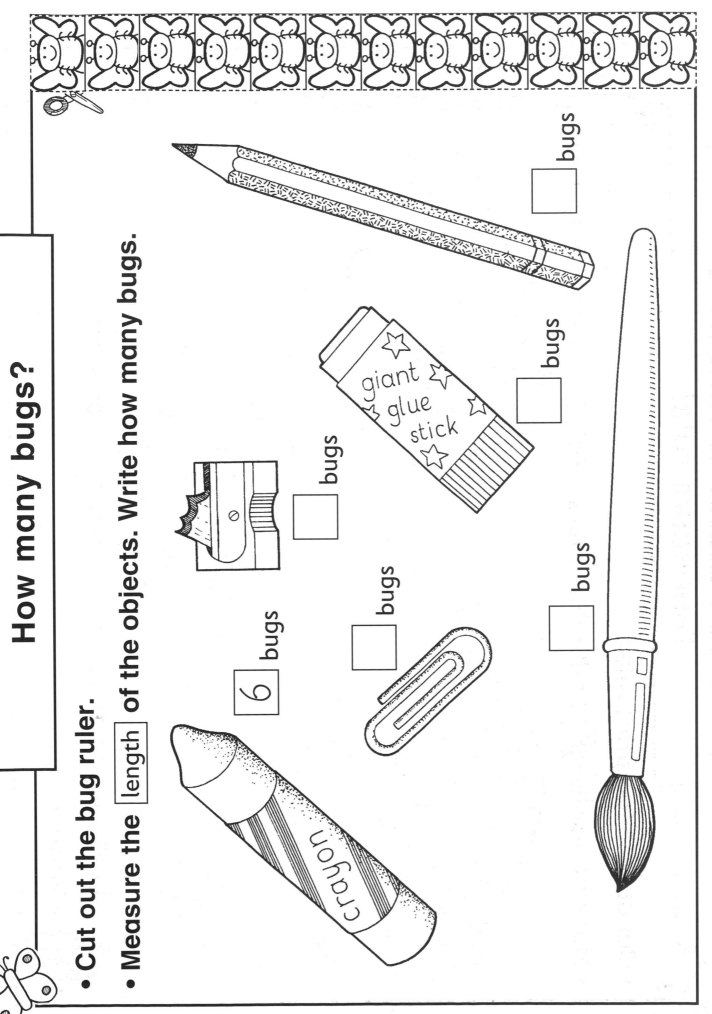

bugs

bugs

bugs

bugs

bugs

bugs

6 bugs

Teachers' note Ensure that the children measure the length of each object, not the height. They could measure the objects on the page and then measure the real objects to compare. Remind the children to line up the edge of the ruler with the edge of the object. Begin to discuss the language of approximation, such as nearly, just over, just under.

Developing Numeracy
Measures, Shape and Space
Year 1
© A & C Black

9

Metre measuring

- **Join the label to the correct object.**

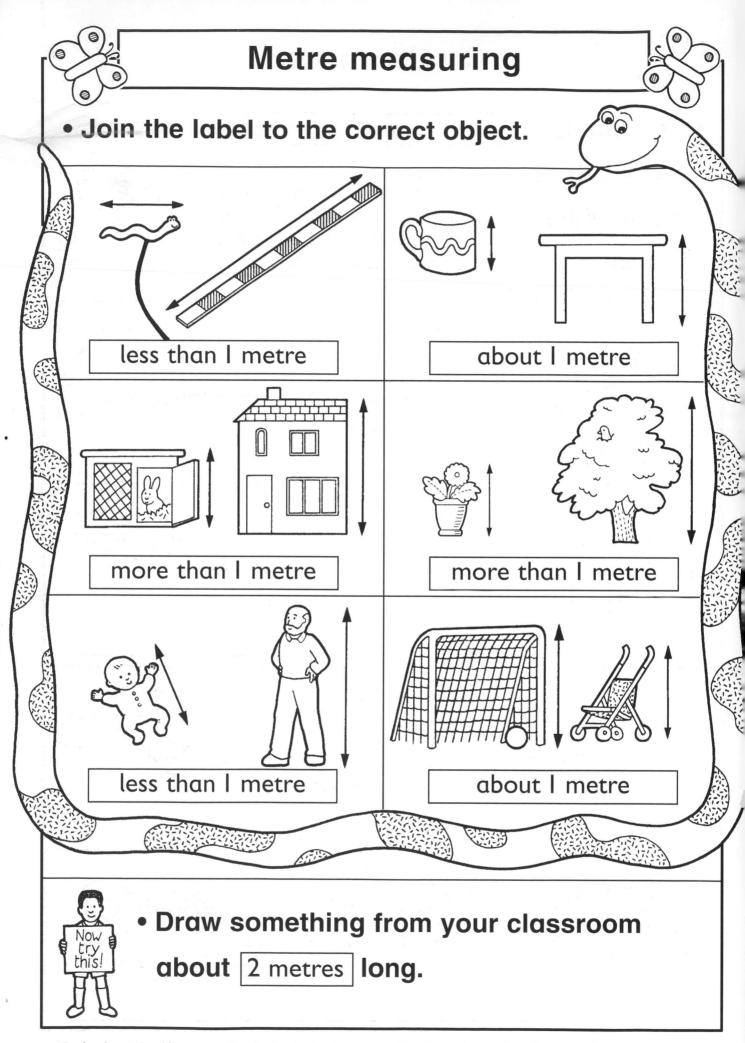

less than I metre

about I metre

more than I metre

more than I metre

less than I metre

about I metre

- **Draw something from your classroom about** 2 metres **long.**

Now try this!

Teachers' note Provide opportunities for the children to practise estimating and measuring using a metre stick. Remind them of the language of approximation. Some children may need a metre stick on the table for reference.

Developing Numeracy
Measures, Shape and Space
Year 1
© A & C Black

• **You can use** ┃centimetres┃ **or** ┃metres┃ **to measure**

these things.

Write which

is best.

Remember, **you** are about
I metre tall, and your finger
is about I centimetre wide.

 centimetres	 _____	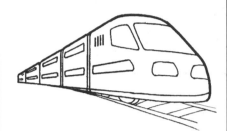 _____
 _____	 _____	 _____
 _____	 _____	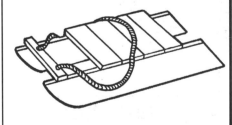 _____
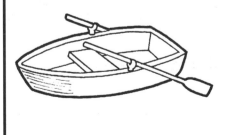 _____	 _____	 _____

Teachers' note Encourage the children to make comparisons between the objects pictured and themselves; ask them which things are longer or taller than themselves. As an extension activity, the children could cut out the cards and sort them into sets using different criteria, for example: smallest to largest; wheels/no wheels; engine/no engine; land/air/sea, and so on.

**Developing Numeracy
Measures, Shape and Space
Year 1**
© A & C Black

Throw the beanbag

The children are throwing beanbags.

They measure their throws in metres .

- **Complete the sentences.**

  Alia threw her beanbag ___3___ metres.

  Claire threw her beanbag _____ metres.

  Bill threw his beanbag _____ metres further than Tom.

  Tom threw his beanbag I metre further than _____.

  _____ threw the shortest distance.

  _____ threw the longest distance.

Teachers' note This is an activity that the children can try for themselves. They can measure either using non-standard units, such as strips of paper, or in metres.

**Developing Numeracy
Measures, Shape and Space
Year 1**
© A & C Black

Animal game

You need:
a dice
a counter each
a balance
some cubes

☆ Take turns to throw the dice. Move your counter.

☆ If you land on an animal, name another animal that is **heavier** or **lighter**.

☆ If your partner agrees, put a cube in your side of the balance.

☆ The winner is the one with the heaviest pile of cubes.

start

heavier

lighter

heavier

heavier

lighter

lighter

lighter

heavier

lighter

lighter

heavier

Teachers' note Discuss positions of a bucket balance to show lighter or heavier, and demonstrate with toys. Brainstorm a list of animals and discuss them using comparisons of weight: heavier than and lighter than. Before beginning the game, decide on the number of circuits to be completed. Encourage the children to choose different animals from each other.

Developing Numeracy
Measures, Shape and Space
Year 1
© A & C Black

13

Heaviest and lightest

These objects are in order,  heaviest to lightest .

• Join the heaviest and lightest objects to their pans.

Teachers' note Discuss the objects in the line and describe them in terms of: 'This is the heaviest/ heavier than…' The activity can be carried out practically, using a bucket balance and a similar set of objects. Remind the children of the process for finding the heaviest or lightest of three objects.

Developing Numeracy
Measures, Shape and Space
Year 1
© A & C Black

Make it balance

You need:
a bucket balance
some cubes
the objects shown

• **Write how many cubes balance each object.**

☐ cubes	☐ cubes
☐ cubes	☐ cubes
☐ cubes	☐ cubes

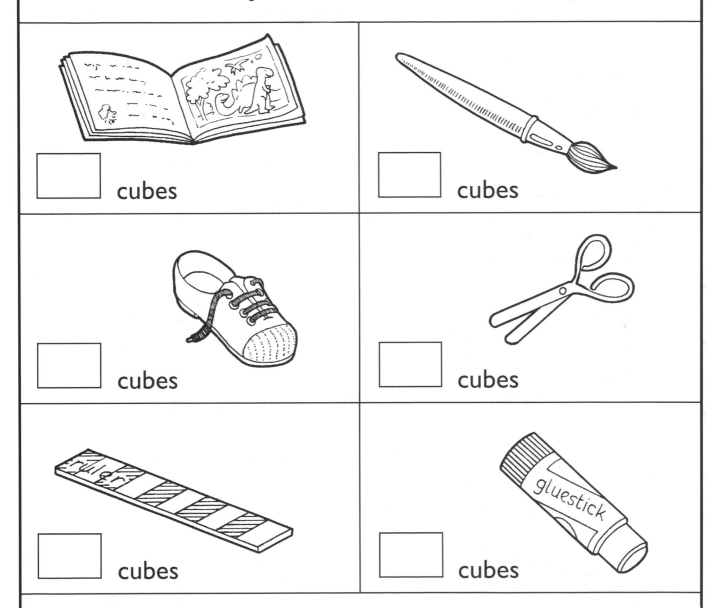

 Now try this!

• **Tick the** heaviest **object.**

• **Find something that you think is heavier.**

• **Check with the balance and cubes.**

Teachers' note This activity can be done in pairs. Each pair will need a bucket balance, some cubes and the items pictured. Ask the children to estimate how many cubes will balance an object, then to check using the bucket balance. Ask them to suggest something heavier than the object and repeat the activity. This can be repeated for something lighter than the object.

Developing Numeracy
Measures, Shape and Space
Year 1
© A & C Black

Fruit balance

You need:
a bucket balance
some cubes
fruit and vegetables

- **Choose a fruit or vegetable to balance with cubes.**

- **Write your** estimate .

- Check **by balancing.**

I chose	I estimated	I measured
	_____ cubes	_____ cubes
	_____ cubes	_____ cubes
	_____ cubes	_____ cubes
	_____ cubes	_____ cubes
	_____ cubes	_____ cubes

The heaviest thing I chose was a _____.

The lightest thing I chose was a _____.

Now try this!

- **Find something in the classroom which balances** 12 **cubes.**

Teachers' note Provide a range of fruit and vegetables. The children should work in pairs and choose one at a time. Encourage them to use the information from their first object to help with their next estimate.

**Developing Numeracy
Measures, Shape and Space
Year 1**
© A & C Black

Comparing toys

10 cubes

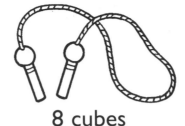
8 cubes

12 cubes

15 cubes

6 cubes

20 cubes

- **Choose an object that is** [heavier] **.**

- **Draw it. Write how much heavier.**

15 cubes   _____ cubes heavier

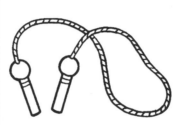

8 cubes   _____ cubes heavier

12 cubes   _____ cubes heavier

10 cubes   _____ cubes heavier

Now try this!

- **Find something in the classroom heavier than** [20] **cubes.**

I found a _____. It weighed _____ cubes.

Teachers' note Ask the children to decide which is the lightest object on the sheet. Repeat for the heaviest. Ask them to describe the other objects in terms of: 'This toy is heavier/lighter than...' The sheet could be made easier or more challenging by changing the number of cubes per toy.

Developing Numeracy
Measures, Shape and Space
Year 1
© A & C Black

Toy balance

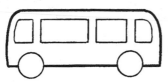

3 cubes · 6 cubes · 4 cubes · 7 cubes

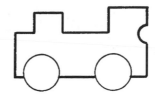

9 cubes · 10 cubes

• Draw 2 toys in the bucket to make it balance.

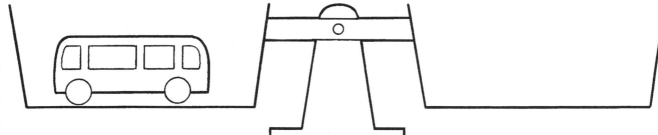

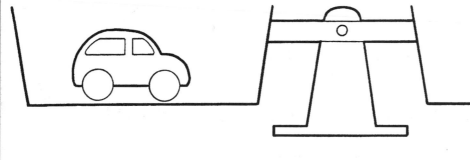

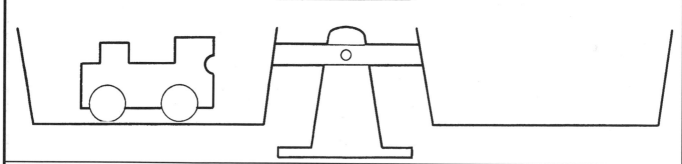

• Draw toys to balance this plane.

12 cubes

Teachers' note Demonstrate using a bucket balance how two objects can balance a third. Weigh the items using cubes and write an addition, for example, 4 cubes + 5 cubes = 9 cubes.

Developing Numeracy
Measures, Shape and Space
Year 1
© A & C Black

At the circus

clown

ringmaster

juggler

clown

juggler

cowboy

Word-bank

clown   ringmaster

cowboy   juggler

- **Use the word-bank to finish the sentences.**

The ___ringmaster___ is the heaviest.

The _____ is the lightest.

The clown is lighter than the _____.

The clown is heavier than the _____

and the _____.

The juggler is lighter than the _____

and the _____.

- **Draw the people in order, lightest first.**

Teachers' note Discuss the pictures carefully, using the language of weight. This activity can be carried out practically using classroom items. The children should compare two items each time and then order them by weight.

Developing Numeracy
Measures, Shape and Space
Year 1
© A & C Black

Drink and shrink!

- **Colour blue the bottle that holds the** most .
- **Colour red the bottle that holds the** least .

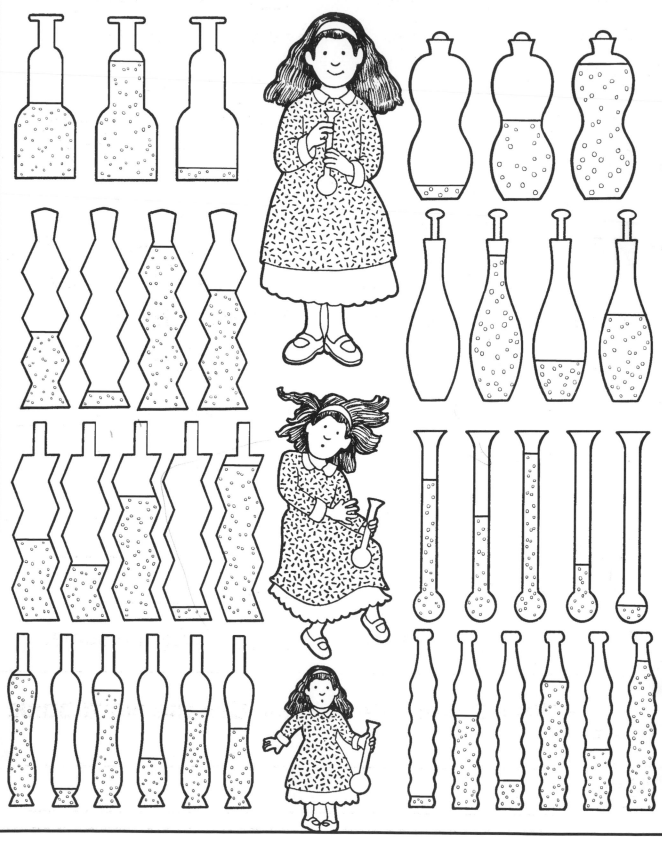

Teachers' note Provide opportunities for the children to make direct comparisons of capacity by filling and pouring.

**Developing Numeracy
Measures, Shape and Space
Year 1**
© A & C Black

Container quiz

8 cups   6 cups   6 cups   10 cups

9 cups   2 cups

Word-bank

more most mug vase bucket

less least jug teapot bottle

- **Finish the sentences. Use the word-bank.**

The bucket holds _____more_____ than the teapot.

The jug holds _____ than the bucket.

The bucket holds the _____.

The mug holds the _____.

The teapot holds _____ than the _____.

The vase holds _____ than the _____.

The _____ and the _____ hold the same.

- **Write 2 more sentences. Use the word-bank.**

_____

_____

Teachers' note Provide opportunities for the children to experience using non-standard units for filling and pouring, using sand, water or lentils. Encourage them to state which container holds more, less, most and least.

Developing Numeracy
Measures, Shape and Space
Year 1
© A & C Black

About the same

- **Look at the containers below.**

- **Choose something from the table that you think holds** about the same **. Draw it.**

holds about the same as

holds about the same as

holds about the same as

holds about the same as

holds about the same as

Teachers' note Before beginning the activity, discuss the containers with the children. They may find it helpful to carry out the activity practically. As an extension activity, the children could find other objects which they think hold about the same as each other and check by filling and pouring.

**Developing Numeracy
Measures, Shape and Space
Year 1**
© A & C Black

Choose a unit

- **Ring the best unit for measuring each container.**

Container	Unit		

- **Draw 2 more containers.**
- **Draw units for measuring them.**

Teachers' note Discuss what makes a unit better for measuring, i.e. not too big and not too small. The children could work in pairs to discuss their ideas. Provide a range of containers for the children to refer to.

**Developing Numeracy
Measures, Shape and Space
Year 1
© A & C Black**

Litre measuring

- **Cut out the cards. Sort them into the pools.**

holds less than a litre	holds about a litre	holds more than a litre

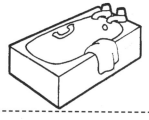

Teachers' note Provide opportunities for the children to measure capacity in litres. Look together at bottles and discuss how '1 litre' is written on the labels. As an extension, the children could list all the containers they can think of that contain less than/more than a litre.

Developing Numeracy
Measures, Shape and Space
Year 1
© A & C Black

Draw a unit

• **Draw the best unit for measuring each container.**

Units

spoon

cup

1 litre jug

 →

unit

 →

unit

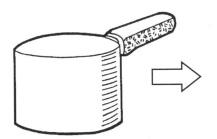 →

unit

  →

unit

  →

unit

 →

unit

• **Draw 2 more containers you could measure with the | 1 litre jug |.**

Teachers' note Provide opportunities for the children to fill containers using different units, including a 1 litre jug. Emphasise the importance of choosing a suitable unit.

Developing Numeracy
Measures, Shape and Space
Year 1
© A & C Black

Changes

- **Join the pictures to the correct** | season | .

- **Write the** | months | **for each season.**

Teachers' note Talk about how the seasons change and show photographs that demonstrate seasonal changes. Check that the children know what sort of clothes would normally be worn for each season. Some children may need a word-bank for the extension activity.

**Developing Numeracy
Measures, Shape and Space
Year 1**
© A & C Black

Birthday cake

- **Cut out the pictures.**

- **Put them in the correct order.**

- **Make your own picture story. Cut it out.**

- **Ask a partner to put the pictures in order.**

Teachers' note Talk about the sequence of events when making a cake. As an extension activity, you could help the children to find cartoons in comics which they can cut up and reorder.

Developing Numeracy
Measures, Shape and Space
Year 1
© A & C Black

27

Time snap: 1

2 o'clock	7 o'clock	8 o'clock	12 o'clock
9 o'clock	1 o'clock	3 o'clock	10 o'clock
11 o'clock	5 o'clock	6 o'clock	4 o'clock

Teachers' note These cards can be used to play Snap or Pelmanism. They can be combined with the 'half past' cards on page 29 or used on their own.

Developing Numeracy
Measures, Shape and Space
Year 1
© A & C Black

Time snap: 2

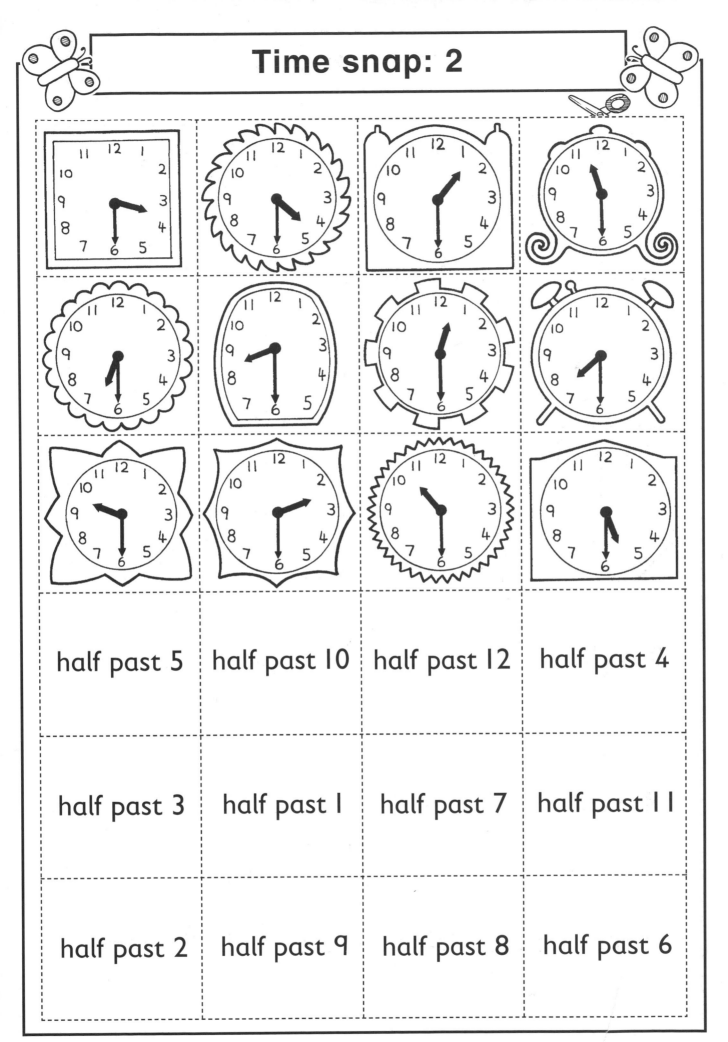

half past 5	half past 10	half past 12	half past 4
half past 3	half past 1	half past 7	half past 11
half past 2	half past 9	half past 8	half past 6

Teachers' note These cards can be used to play Snap or Pelmanism. They can be combined with the 'o'clock' cards on page 28 or used on their own.

Developing Numeracy Measures, Shape and Space Year 1 © A & C Black

Sam's day

- ## Cut out the pictures of Sam's day.

- ## Put them in the correct order.

- ## Draw 3 pictures of <u>your</u> day.

- ## Draw a clock on each one to show the time.

- ## Cut out the pictures.

- ## Ask a partner to put them in order.

Teachers' note Discuss with the children the day's events and identify the time for each one. Ask the children to sequence the events by putting them in time order. Encourage them to use the language of time: before, next, after and so on.

Developing Numeracy Measures, Shape and Space Year 1 © A & C Black

My day

• **Join the clocks to the correct labels.**

get up	go home
eat breakfast	teatime
go to school	have a bath
lunchtime	go to bed

• **Draw the clock hands for your favourite time of day.**

• **Write why it is your favourite.**

Teachers' note Discuss what the time is when the children carry out the activities listed. Talk about how on some days they may get up earlier or later, and so on.

**Developing Numeracy
Measures, Shape and Space
Year 1**
© A & C Black

Just a minute

- **What do you think will take about** $\boxed{\text{I minute}}$ **?**
- **Colour the label.**

building a tower	eating a meal	playing skittles
making a cake	writing your name	walking to shops
brushing your teeth	washing up	fastening your shoes
reading a book	drinking squash	painting a picture

- **Try doing some of the things.**
- **Time yourself with a I minute sand timer.**

Teachers' note Ask the children to shut their eyes for what they think is a minute. Then discuss how accurate their estimations were. Provide opportunities for the children to do things for a minute. Let them find out how many times can they draw a square or write a word.

**Developing Numeracy
Measures, Shape and Space
Year 1
© A & C Black**

Matti's week

Monday	Tuesday	Wednesday	Thursday
made a cake	played tennis	visited Gran	bought a scooter

Friday	Saturday	Sunday
watched a video	went to a birthday party	went to the seaside

- **On which day did Matti**

  buy a scooter? ___Thursday___

  go to the seaside? _____

  eat birthday cake? _____

  play tennis? _____

- **What did Matti do**

  on Monday? _____

  the day after Thursday? _____

  the day before Thursday? _____

- **Write 2 more questions about Matti.**
- **Ask a partner to answer them.**

Teachers' note Encourage the children to recite the days of the week in order and to describe the week using language such as 'today', 'tomorrow' and 'yesterday'.

**Developing Numeracy
Measures, Shape and Space
Year 1
© A & C Black**

What's the shape?

• **Join the objects to the labels.**

The labels can have more than one object.

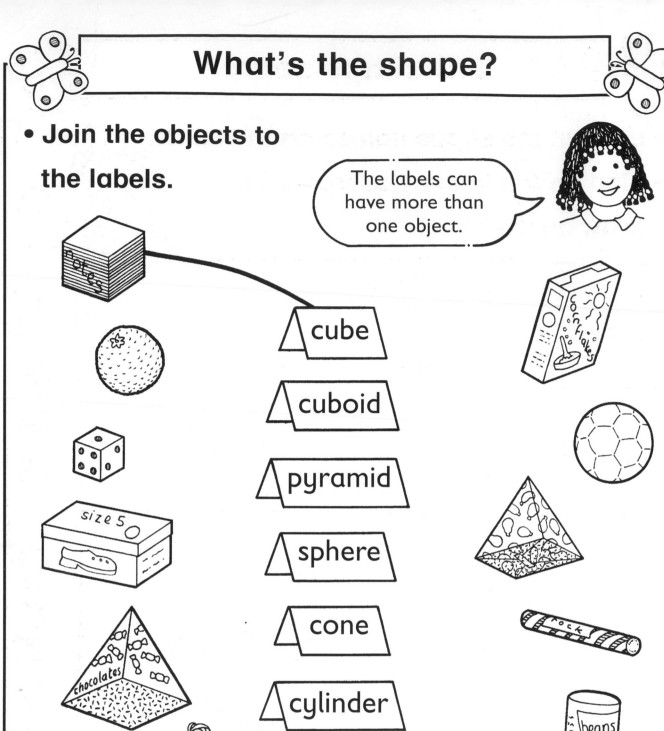

cube

cuboid

pyramid

sphere

cone

cylinder

• **Draw two more** cuboid **objects.**

Teachers' note Before beginning the activity, revise the names of the shapes and introduce the terms 'cuboid' and 'cylinder'. Encourage the children to identify everyday objects by shape. You could play the 'Behind the wall shapes' game from page 5.

Developing Numeracy
Measures, Shape and Space
Year 1
© A & C Black

3-D shape search

- **Look at the shape names on the chart.**
- **Find things in the classroom to draw on the chart.**

cube	
cuboid	
cone	
cylinder	
pyramid	
sphere	

Teachers' note Encourage the children to describe the things they find by their mathematical properties and names.

Developing Numeracy Measures, Shape and Space Year 1 © A & C Black

Sand shapes

- **Rupa is pressing shapes into wet sand.**
- **Colour the** faces **that belong to each shape.**

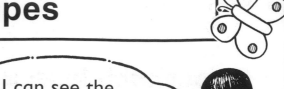

I can see the shape of the **face** in the sand.

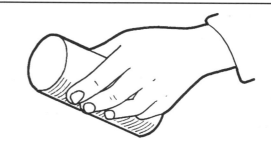

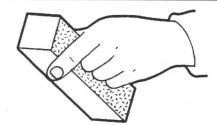

- **Which shapes have** square **faces?**
- **Say their names.**

Teachers' note Encourage the children to identify 3-D shapes by the shapes of their faces. You may wish to provide real 3-D shapes to support the children when completing this activity.

**Developing Numeracy
Measures, Shape and Space
Year 1
© A & C Black**

Busy street

- **Colour each label a different colour.**

| sphere | cylinder | cone | cube | pyramid | cuboid |

- **Colour the shapes to match the labels.**

Teachers' note The children will need pencil crayons in six different colours. Some children may require a visual reminder of the 3-D shapes and their names.

**Developing Numeracy
Measures, Shape and Space
Year 1
© A & C Black**

37

Shape riddles

• **Join the riddles to the shapes.**

I have 6 faces all the same.

I have 5 corners.

I have 6 faces. Only 4 are the same.

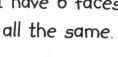

I have 1 curved face.

I have 1 face that is a circle.

I have 2 faces that are circles.

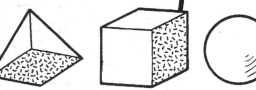

• **Choose 2 of the shapes.**

• **Write another riddle for each one.**

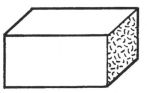

Join your riddle to the shape.

Now try this!

Teachers' note Before beginning this activity, discuss which 3-D shapes have flat/curved faces. Encourage the children to describe the shapes by their properties. Some children may need real 3-D shapes for reference.

Developing Numeracy
Measures, Shape and Space
Year 1
© A & C Black

Face race

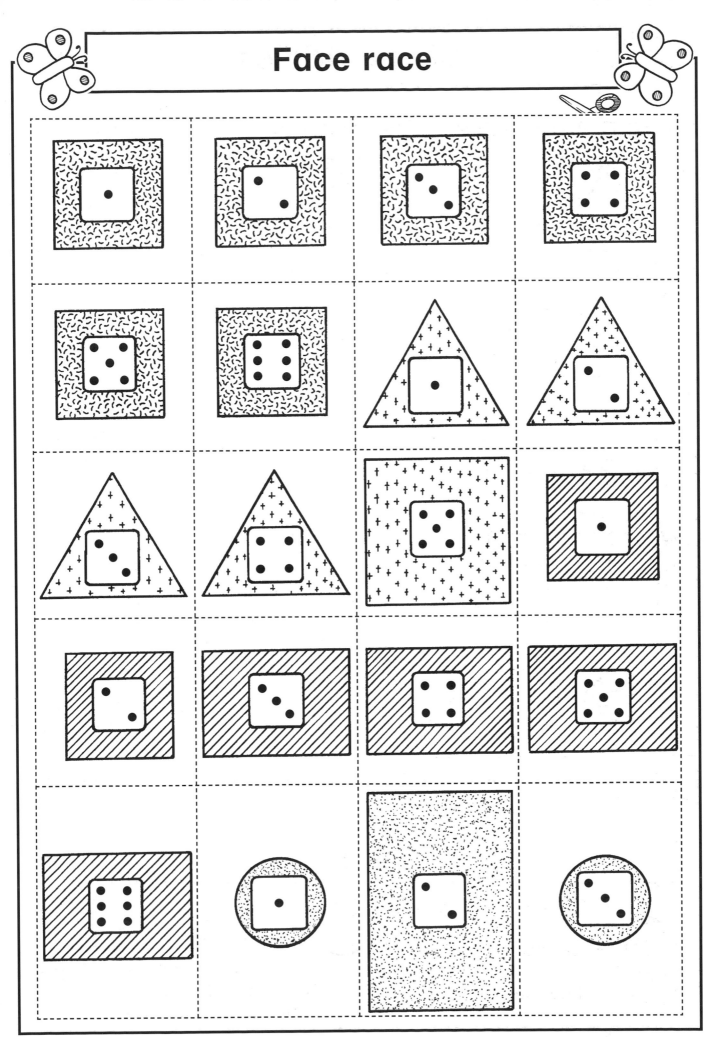

Teachers' note Provide a copy of the sheet for each child. Cut out the cards and spread them face up. The children take turns, in small groups, to roll a dice and pick a card that matches the number rolled. If there is no card to match the dice roll, they miss a go. The winner is the first to collect all the faces for one shape (cube, pyramid, cuboid or cylinder).

Developing Numeracy
Measures, Shape and Space
Year 1
© A & C Black

Make a pattern

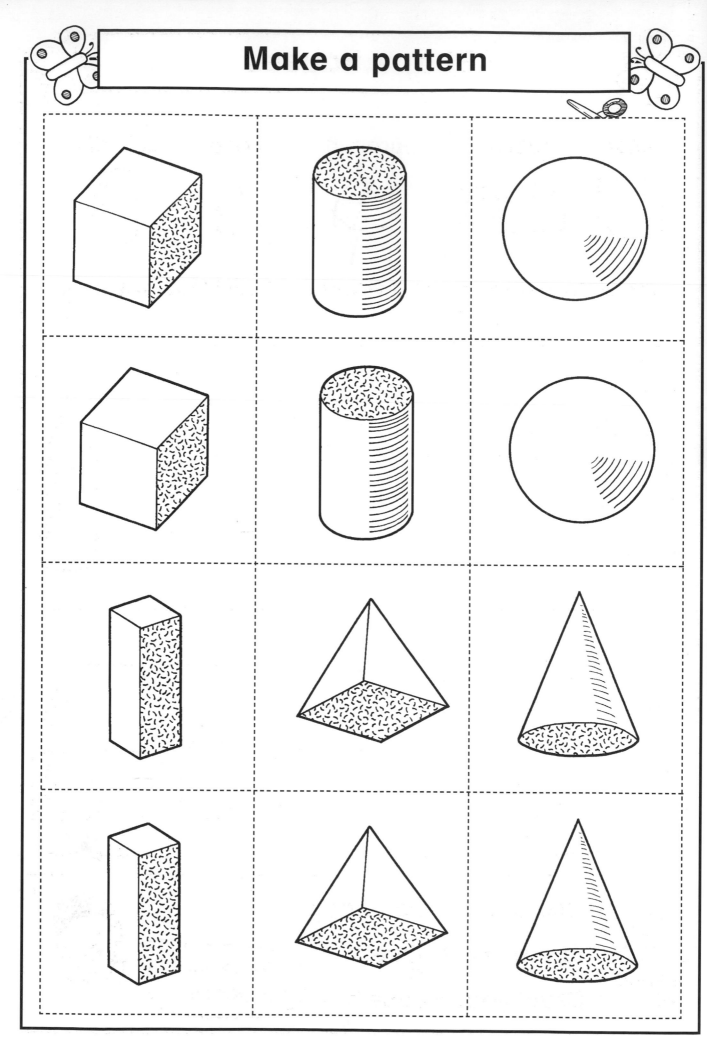

Teachers' note Provide a copy of the sheet for each child. Cut out the cards. The children work in pairs: one child begins a repeating pattern using the cards, and the other child continues the pattern. They could then copy their pattern using real 3-D shapes. Let the children swap roles and repeat the activity.

**Developing Numeracy
Measures, Shape and Space
Year 1**
© A & C Black

Shape houses

cube   cuboid   pyramid   cone   cylinder

- **Write the shapes you need to build the houses.**

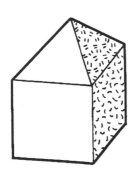 pyramid _____

_____

 _____

_____

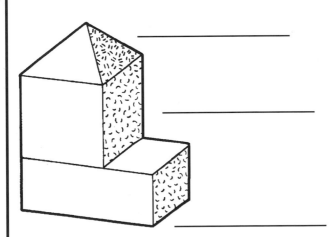 _____

_____

_____

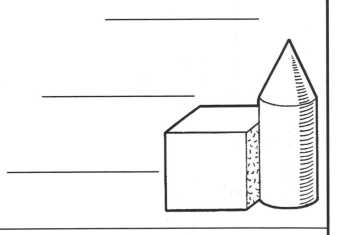 _____

_____

_____

- **Use one of each shape to make a house.**
- **Draw your house.**

Work with a partner.

Teachers' note Check that the children can name the 3-D shapes and can recognise them from the drawings. Discuss each shape's properties, naming faces and talking about flat/curved faces.

**Developing Numeracy
Measures, Shape and Space
Year 1**
© A & C Black

41

Draw the shape

- **Look at the shapes.**

- **Draw them in the correct places.**

Copy the shapes carefully.

circle	rectangle
square	triangle

- **Take turns with a partner.**
- **Describe a shape.**
- **Ask your partner to draw it.**
- **Check it is correct.**

Teachers' note Revise the names of the shapes. Encourage the children to recognise and describe each shape by its properties. You could play the 'Behind the wall shapes' game from page 5.

Developing Numeracy
Measures, Shape and Space
Year 1
© A & C Black

Friendly giant

- **Colour each label a different colour.**

| square | rectangle | triangle | circle |

- **Colour the shapes to match the labels.**

Teachers' note The children will need pencil crayons in four different colours. Before beginning the activity, you could show the children an A3 enlargement of this sheet and point to different shapes to check that the children can name them. As an extension, challenge the children to draw their own shape giant and colour the shapes to match the labels.

**Developing Numeracy
Measures, Shape and Space
Year 1**
© A & C Black

2-D shape search

- **Look at the shape names on the chart.**
- **Find things in the classroom to draw on the chart.**

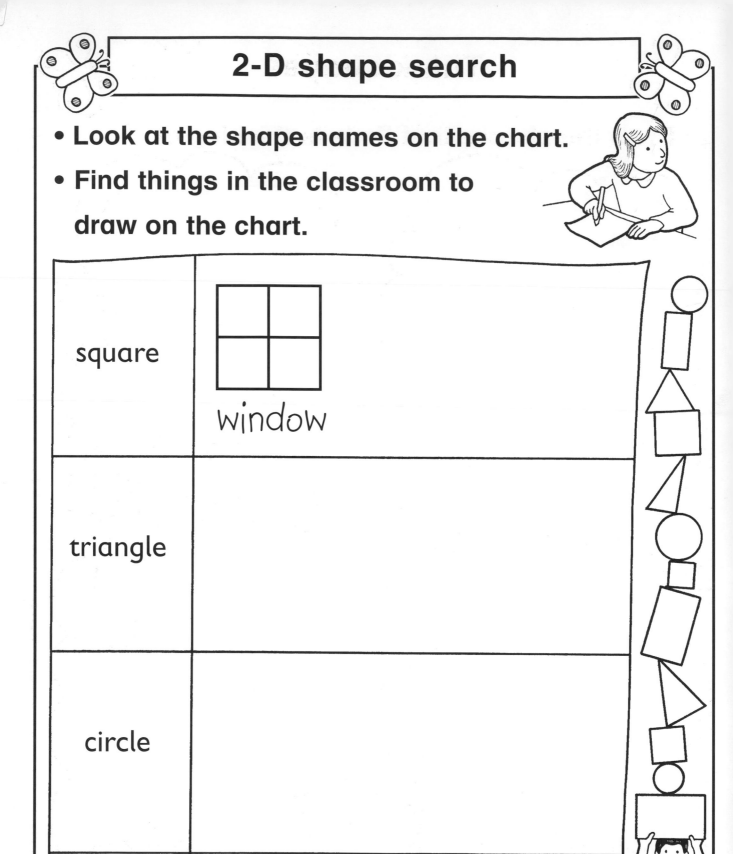

square	window
triangle	
circle	
rectangle	

Developing Numeracy
Measures, Shape and Space
Year 1
© A & C Black

Think shapes

• **Draw the shapes that Jade is thinking of.**

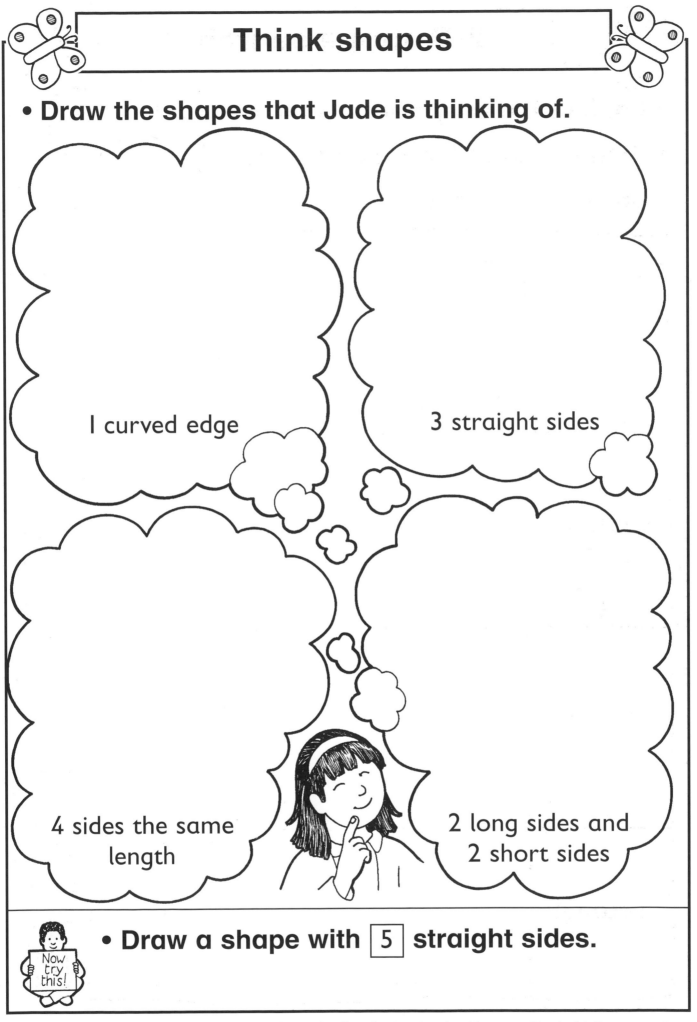

I curved edge

3 straight sides

4 sides the same length

2 long sides and 2 short sides

• **Draw a shape with** 5 **straight sides.**

Now try this!

Teachers' note Remind the children of the vocabulary to describe a shape's properties: side, edge, straight, curved, corners, points.

**Developing Numeracy
Measures, Shape and Space
Year 1
© A & C Black**

Shape train

- **Take turns to read the words inside a carriage.**
- **Find the shape.**
- **Cover the shape with a counter.**

Play with a partner.

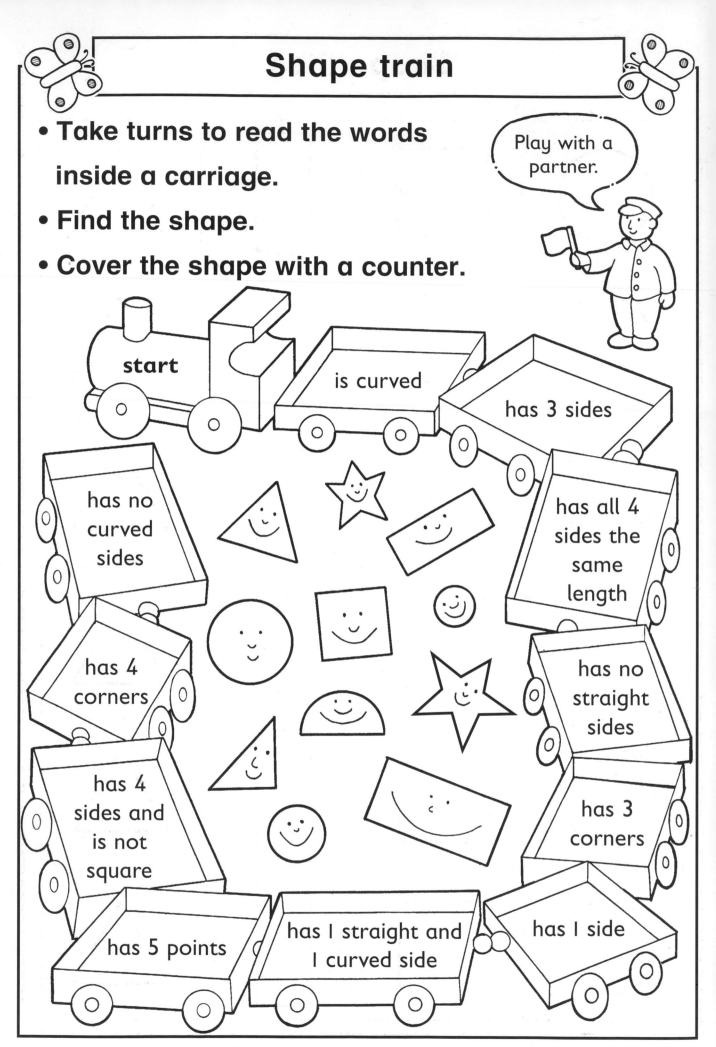

start

is curved

has 3 sides

has no curved sides

has all 4 sides the same length

has 4 corners

has no straight sides

has 4 sides and is not square

has 3 corners

has 5 points

has 1 straight and 1 curved side

has 1 side

Teachers' note Each pair of children will need 11 counters. Explain that they should follow the carriages in sequence. You could ask them to describe each of the shapes by its properties, such as number of sides, curved/straight sides, number of corners/points. Some children may need help reading the descriptions before they begin the activity.

Developing Numeracy
Measures, Shape and Space
Year 1
© A & C Black

Shape tiles

Teachers' note Cut out the shape tiles and use them with the activities on pages 48 and 49.
Alternatively, the children could use the tiles to play Snap or Pelmanism.

Developing Numeracy
Measures, Shape and Space
Year 1
© A & C Black

Shape patterns

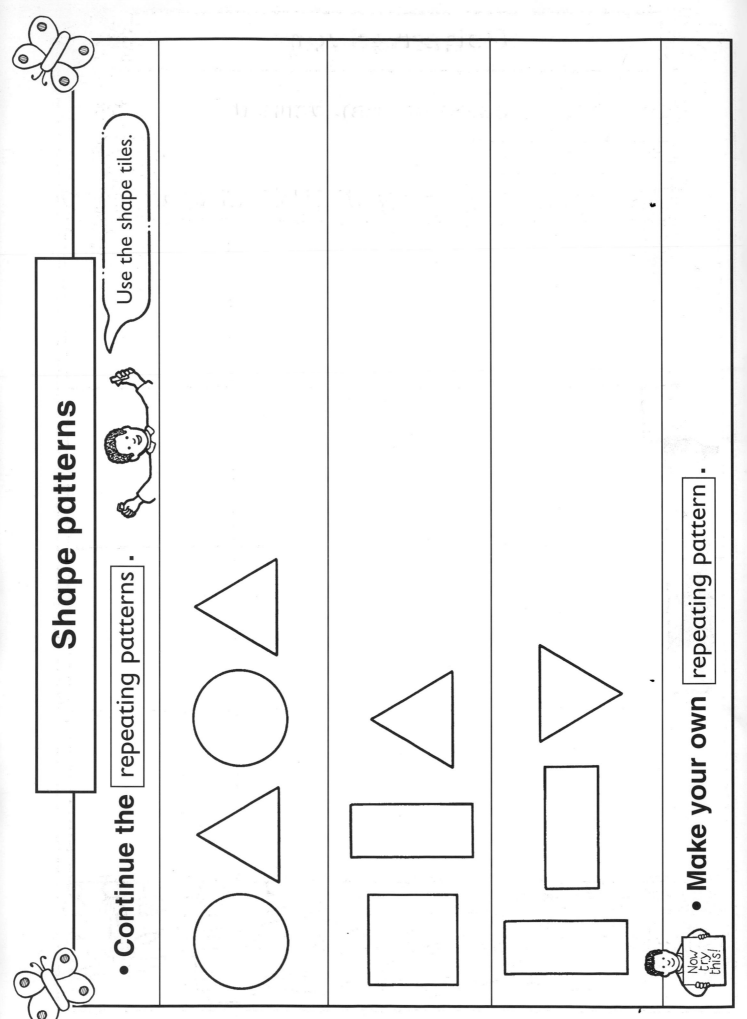

•Continue the repeating patterns.

Use the shape tiles.

•Make your own repeating pattern.

Now try this!

Teachers' note The children will need copies of the shape tiles on page 47. Check that they understand what is meant by a repeating pattern. Encourage them to name the shapes in the patterns and to state what will come next.

**Developing Numeracy
Measures, Shape and Space
Year 1**
© A & C Black

Patterned rug

- **Place shape tiles on the rug. Make a** repeating pattern .

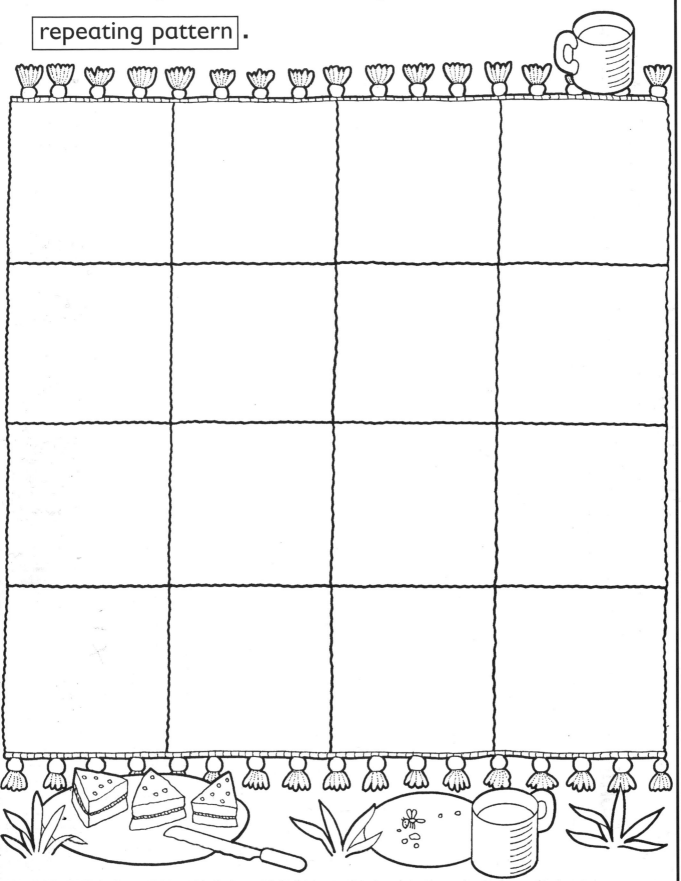

Teachers' note The children will need copies of the shape tiles on page 47. Check that they understand what is meant by a repeating pattern. Encourage them to say their patterns by naming the shapes in order. When they are happy with their patterns they can stick them down.

**Developing Numeracy
Measures, Shape and Space
Year 1**
© A & C Black

Fold the picture

- **Cut out the cards.**
- **Fold each picture in** half .

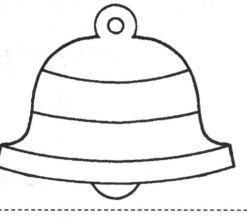

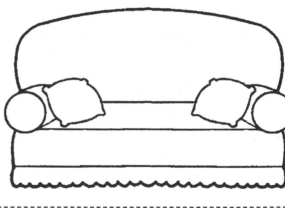

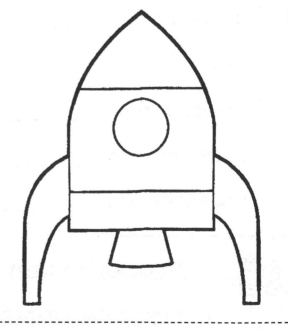

Teachers' note If the children have difficulty with this activity, provide mirrors and ask them to place the mirror on the line of symmetry. The children should discuss how the two sides of each picture match. Ask: 'What is the same?' 'Is anything different?'

Developing Numeracy
Measures, Shape and Space
Year 1
© A & C Black

Where are the toys?

Use the word-bank to help you.

Word-bank

on
under
next to
above

• **Finish the sentences.**

The teddy is _____on_____ the table.

The car is _____ the table.

The balloon is _____ the table.

The bricks are _____ the teddy.

The skates are _____ the floor.

• **Draw a ball on the picture.**

• **Describe where it is.**

Teachers' note Introduce the activity by asking the children to describe the location of something in the classroom, using the language of position.

Developing Numeracy
Measures, Shape and Space
Year 1
© A & C Black

At the seaside

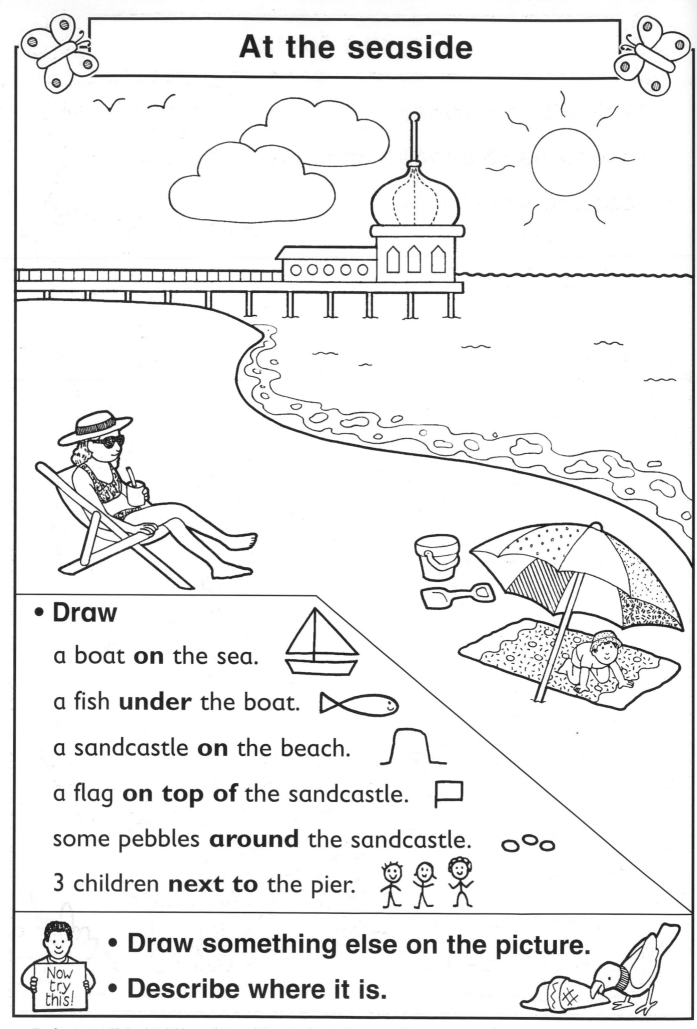

- **Draw**

  a boat **on** the sea.

  a fish **under** the boat.

  a sandcastle **on** the beach.

  a flag **on top of** the sandcastle.

  some pebbles **around** the sandcastle.

  3 children **next to** the pier.

- **Draw something else on the picture.**
- **Describe where it is.**

Now try this!

Teachers' note Show the children a picture from a big book. Ask them to describe what they can see, using the language of position. You may wish to read the instructions with the children before they begin the activity.

Developing Numeracy
Measures, Shape and Space
Year 1
© A & C Black

On the farm

- **Work with a partner.**
- **Take turns to choose an animal.**
- **Describe where it is.**

Use the word-bank to help you.

Word-bank

over	under
above	below
beside	next to
between	on in
in front of	behind

Teachers' note Encourage the children to use the language of position to describe things they can see in the classroom. This could be an I Spy game: 'I spy something on top of the cupboard… under the table…' and so on. Ask the children to read aloud the words in the pond.

**Developing Numeracy
Measures, Shape and Space
Year 1**
© A & C Black

Moving in

- **Draw**

  a TV **beside** the fire.

  a chair **underneath** the window.

  a picture **above** the fire.

  a rug **in the centre of** the room.

  a cat **in front of** the fire.

  a dog **between** the rug and the door.

- **Draw something else in the room.**
- **Write where it is.**

Teachers' note You may wish to read the instructions with the children before they begin the activity. Some children may benefit from working with an adult who gives instructions for positioning real objects, for example, in a room of a doll's house.

Developing Numeracy
Measures, Shape and Space
Year 1
© A & C Black

Pixies are go!

- **Draw the pixie planes where they land.**

Pixie 1 flies

up 2 clouds

right 1 cloud

Pixie 2 flies

up 3 clouds

right 1 cloud

down 2 clouds

Pixie 3 flies

up 1 cloud

left 2 clouds

up 2 clouds

- **Captain Pixie visits all the clouds which have planes.**

- **Describe her route.**

Now try this!

Work with a partner.

Teachers' note Revise the directional words 'up', 'down', 'left' and 'right'. Help the children to remember 'left' by showing them how to make an L-shape with their left thumb and fingers.

Developing Numeracy Measures, Shape and Space Year 1
© A & C Black

Mole's day out

- **Draw a way out for Mole.**
- **Describe it to a partner.**

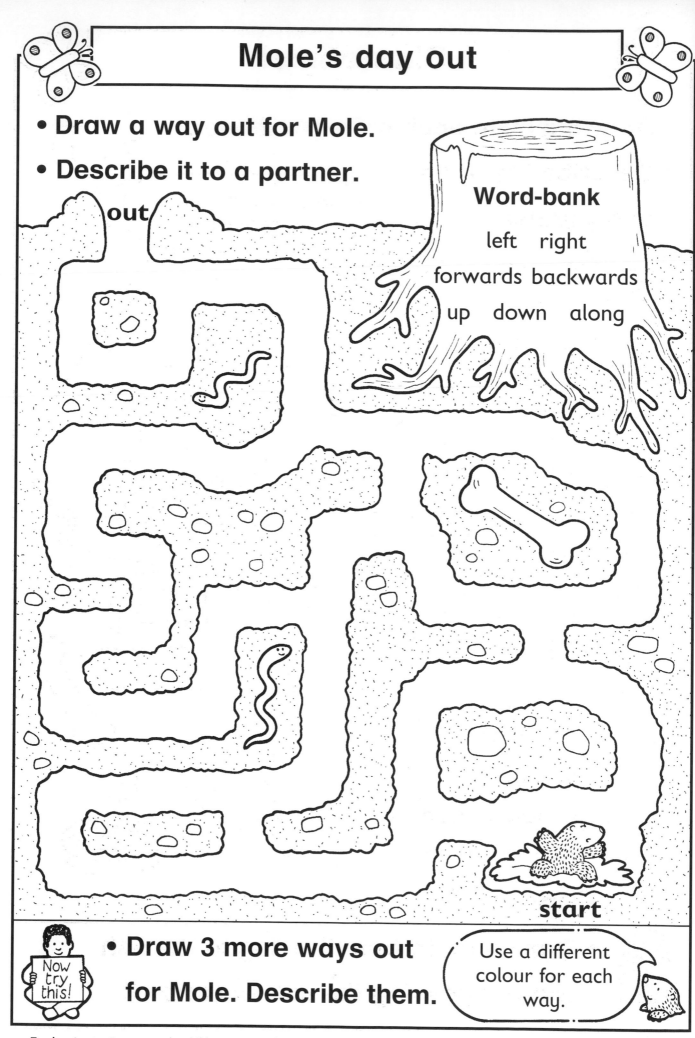

Word-bank

left   right

forwards   backwards

up   down   along

out

start

- **Draw 3 more ways out for Mole. Describe them.**

Now try this!

Use a different colour for each way.

Teachers' note Encourage the children to trace the route through the maze with a finger before they draw it. Ask them to describe the route using the word-bank. If necessary, revise 'left' and 'right'.

Developing Numeracy
Measures, Shape and Space
Year 1
© A & C Black

Dinnertime!

- **Draw a way to each animal's dinner.**

- **Describe the routes**

  to a partner.

Use a different colour for each animal.

Word-bank

left right turn

along past around

between forwards

- **Draw a route from the dog to the fish**

  without passing the cat. Describe it.

Teachers' note Encourage the children to trace the route with a finger before they draw it. Ask them to describe the route using the language of direction. Challenge them to find and describe different routes. Alternatively, the children could work in pairs with a sheet each and a screen (such as a book) between them. One child draws the route and describes it for the other to draw.

Developing Numeracy
Measures, Shape and Space
Year 1
© A & C Black

Amazing escape

- **Put a counter in the centre of the maze.**

- **Tell your partner how to escape.**

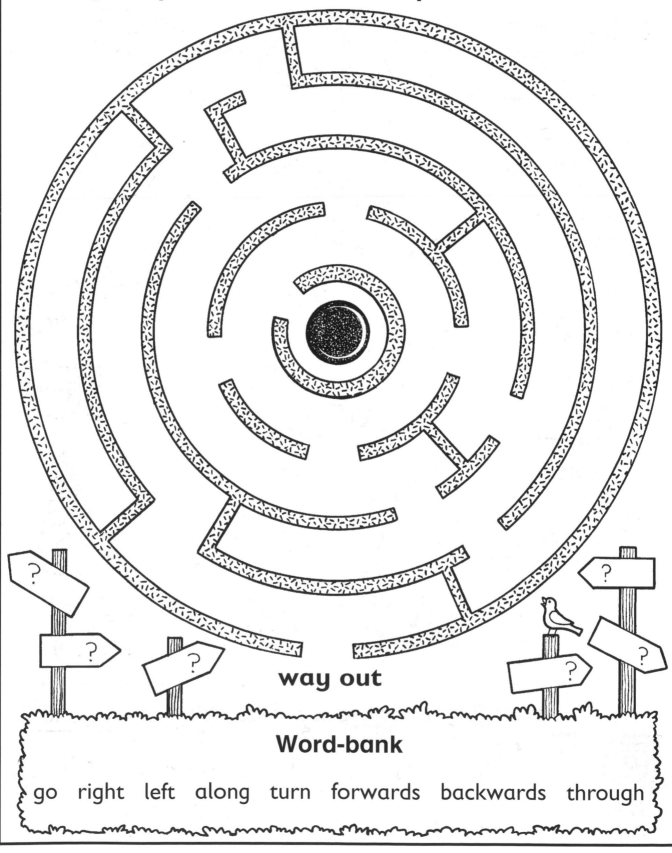

way out

Word-bank

go   right   left   along   turn   forwards   backwards   through

Teachers' note The children should work in pairs. One child instructs the other how to move the counter from the centre to reach the way out. Encourage them to use the language of direction (provided in the word-bank). Pairs of children should sit side by side so that they have the same orientation.

**Developing Numeracy
Measures, Shape and Space
Year 1**
© A & C Black

Helter skelter

Which objects will roll **,** slide **, or** roll <u>and</u> slide **?**

• Join the objects to the correct labels.

The labels can have more than one object.

roll

slide

roll and slide

• Draw 2 more objects that will roll <u>and</u> slide **.**

Now try this!

Teachers' note The children may find it helpful to have real objects and a sloping surface to experiment with practically. Discuss their answers in the plenary.

Developing Numeracy
Measures, Shape and Space
Year 1
© A & C Black

Spinning around

• **Ring the things that turn on a point.**

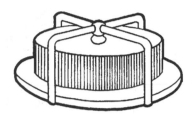  

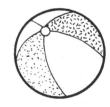 

 • **Find things in the classroom that turn on a point.**
• **Draw them.**

Teachers' note Before beginning the activity, show the children examples of real objects that turn on a point, such as sails on a windmill. Talk about how they turn.

**Developing Numeracy
Measures, Shape and Space
Year 1
© A & C Black**

Whole and half turns

- **Tick the clocks that show a boxed[whole turn].**

- **Cross the clocks that show a boxed[half turn].**

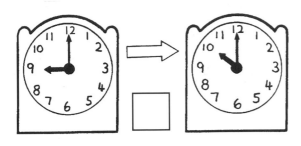

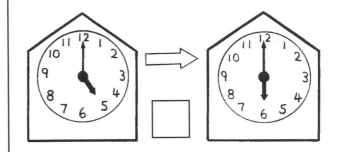

- **Set the hands of a clock to an o'clock time.**

- **Say boxed[whole turn] or boxed[half turn].**

- **Your partner says the new time.**

Take turns with a partner.

Teachers' note To introduce the activity, use a teaching clock to demonstrate whole and half turns. Encourage the children to say the time before and after the hands are turned. Provide small clocks with moveable hands for the extension activity.

Developing Numeracy
Measures, Shape and Space
Year 1
© A & C Black

Turning hands

- ## Join the clock to the new time.

I whole turn

2 whole turns

I half turn

I half turn

I and a half turns

I and a half turns

Teachers' note The children may find it helpful to have a clock with moveable hands so that they can set the hands to the time and then turn them on to the new time. Demonstrate the turns of the hands with a teaching clock.

Developing Numeracy
Measures, Shape and Space
Year 1
© A & C Black

Square puzzle: 1

- **Cut out the square at the bottom of the page.**

- **Cut it into 2 pieces along the dotted line.**

- **Use the pieces to make these shapes.**

- **Use the pieces to make 2 new shapes.**
- **Draw them.**

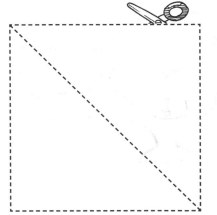

Teachers' note This is a very simple tangram puzzle. Show the children how to turn the pieces until they fit the outlines on the page. Encourage them to re-make the square after making each shape, so that they can see the transformation each time.

Developing Numeracy
Measures, Shape and Space
Year 1
© A & C Black

Square puzzle: 2

- **Cut out the square at the bottom of the page.**

- **Cut it into 4 pieces along the dotted lines.**

- **Use the pieces to make these shapes.**

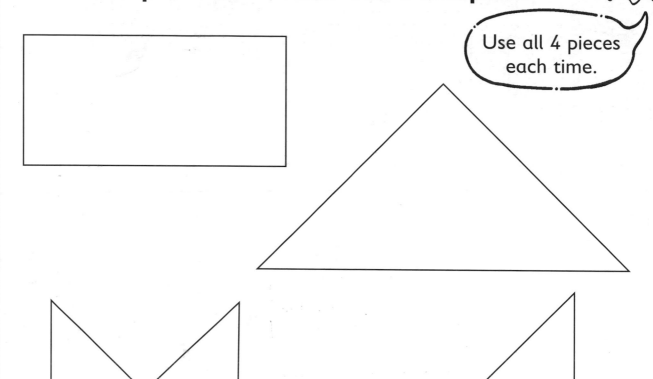

Use all 4 pieces each time.

Now try this!

- **Use the pieces to make 2 new shapes.**

- **Draw them.**

Teachers' note This is a simple tangram puzzle. When the children have cut out the pieces, ask them to re-make the square, then move the pieces from the square to make the first shape. Talk about how the pieces have turned to make the new shape.

Developing Numeracy Measures, Shape and Space Year 1 © A & C Black